How Long?

Contents

Fish

How long is the fish?

The fish is about
four cubes long.

Toy

How long is the toy?

The toy is about
six cubes long.

Scissors

How long are the scissors?

The scissors are about seven cubes long.

Pencil

How long is the pencil?

The pencil is about
eight cubes long.

Glue Stick

How long is the
glue stick?

The glue stick is about six cubes long.

How long are these?

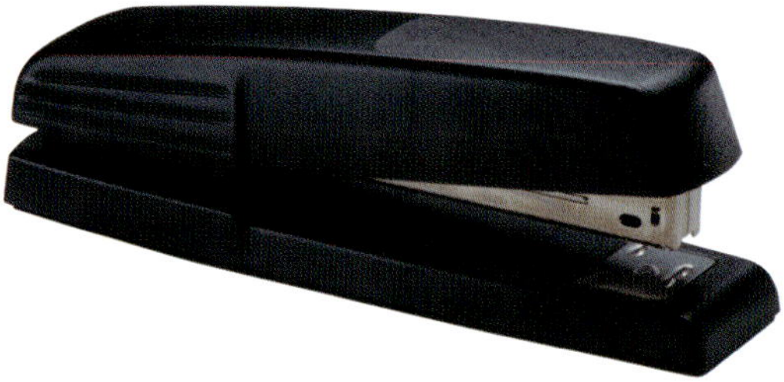